The Oyster

OR, RADIAL SUPPLENESS

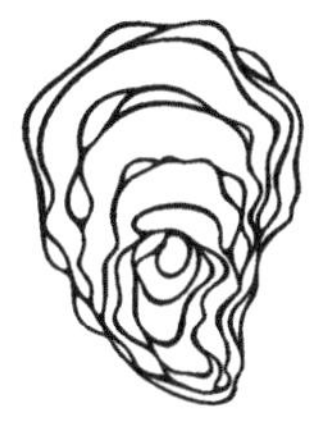

Dejan Lukić and Nik Kosieradzki

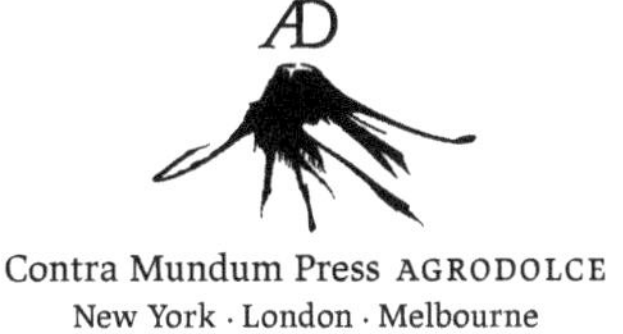

Contra Mundum Press AGRODOLCE
New York · London · Melbourne

First Contra Mundum Press edition 2020.

Library of Congress Cataloguing-in-Publication Data

Lukić, Dejan, 1973–; Kosieradzki, Nik, 1992–

The Oyster. Or, Radial Suppleness / by Dejan Lukić & Nik Kosieradzki

—1st Contra Mundum Press Edition

66 pp., 5×8 in.

ISBN 9781940625423

I. Lukić, Dejan; Kosieradzki, Nik.
II. Title.

2020942532

Contents

The Oyster

OR, RADIAL SUPPLENESS

WITH SHADOW SIDENOTES,
CRYPTIC YET REVEALING A SECRET MOTOR,
A WHISPER, THE INNER THOUGHT THAT FUELS
THE MAIN BODY, I.E. THE WRITING,
LIKE THE OCCASIONAL GRAINS OF SAND
IN THE OYSTER,

DIVIDED IN EIGHT SECTIONS, CURVATURES,
WITH SEVEN IMAGES,
CROSSCUTTING THE REALMS OF THE HUMAN & INHUMAN,
ORGANIC AND INORGANIC,
TASTEFULNESS AND TASTELESSNESS,

WHEREBY A THEORY OF
AMOROUS SUPPLENESS IS ANNOUNCED,
ACCOMPANIED BY THE GEOMETRY OF
FILTERING, EATING, AND PROPAGATING.

AMOR GENERARE.

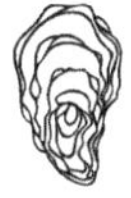

Cleave

There is geometry at the core of formlessness. For neither form nor geometry are stable categories. They constantly renew and vitiate. And life itself can be said to be the rhythmic overflowing that constantly undermines itself (what we call death by dissipation, or eating). Planetary materialism is based on grinding, chewing, slurping, digesting. In that respect the cosmos corresponds to a culinary materialism, operating like a kitchen, and our planet as a cooking pot: here crystals boil (volcanœs), vapors rise (clouds), heat distributes (from tropical to higher latitudes). Our world is filled with sieves, strainers, skimmers, and filters. One such object is the oyster. A fleshy slime, a superior filtering instrument which teaches us a very special geometry of creases and wrinkles that form its shell, the mantle, and soft insides. Both food and filter, the oyster is raw life.

Asteroids, tectonics, teeth;
their chafing creates needed cleft. →

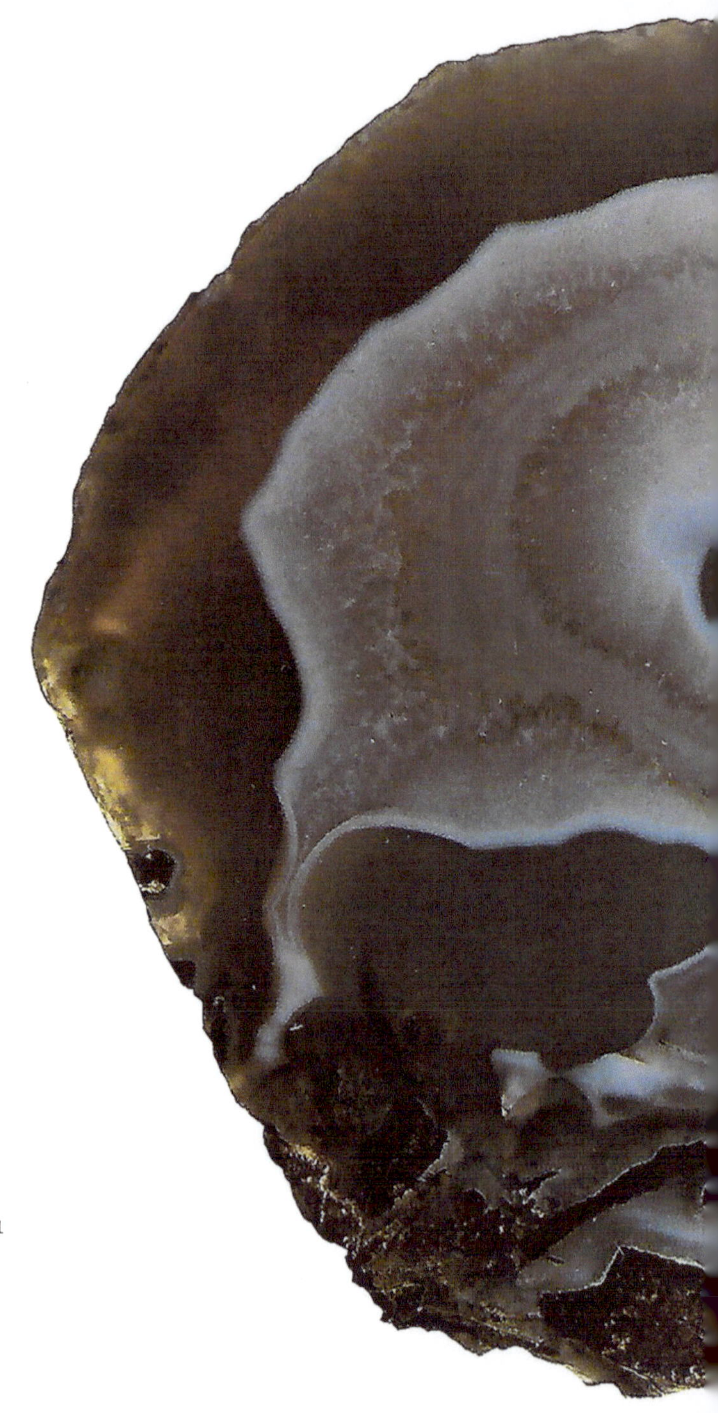

Fig. 1

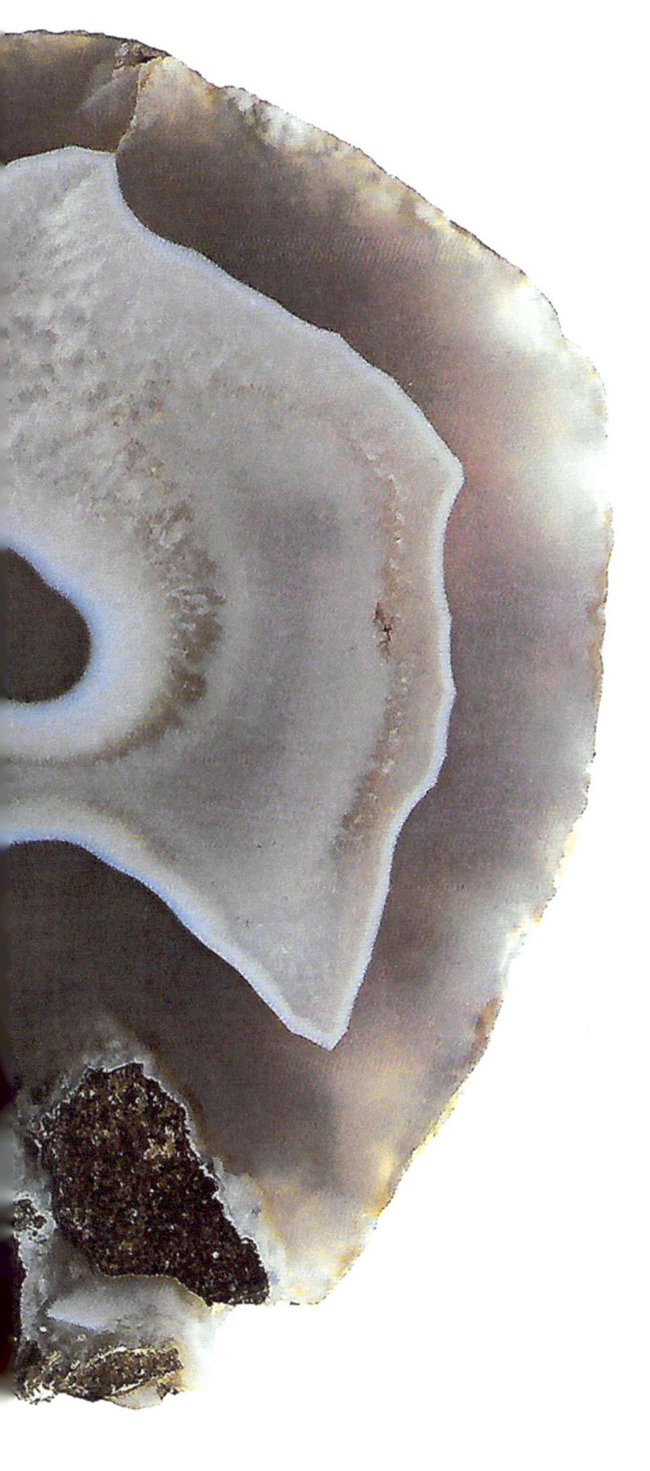

1. Sharp Mineralogy

The oyster is therefore part living, part non-living thing. It attaches to a rock, resembles a rock, and sits opening and closing, folding and unfolding, waiting for nutrients to pass through it. It is alive but it does not feel, it simply exists and filters: a primal life form. But the oyster is never alone; as a multitude it forms large, jagged, ornate colonies, or "beds." This complex outward conglomeration of the thousands resembles the intricate ornamentation of a baroque cathedral.

Yet, sentient matter. →

Its determinedly hard exterior encloses its vulnerable soft interior. Given this closed nature, the inside, upon closer look, gives the appearance of a sacristy in a baroque church (while the furrowed outside resembles a rock washed ashore).[1] For a sacristy contains vestments, vessels, furnishings; in other words, mantles of the sacred. The oyster is such a sacristicial space. It is impermeable except for a single crack that is often hidden amongst a series of other cracks that follow the contours of the true opening, a concealed line of entry. The outside is mineral, the inside organic; the outside rough and folded, the inside bone smooth.

Religion of the inhuman. →

Moreover, the word oyster is related to the Greek word *osteon*, meaning bone. But in Slavic languages it also carries the meaning of "being sharp (*oštar*)." Sharpness and hardness are just two characteristics that bring oysters into the family of stones, where mineralogy reveals mutual affinities between different modalities of existence.

Sometimes we see the stone dreaming of being an oyster (Fig. 1). Here the mineral art of stones conjures, when split in half, the image of the oyster's mantle and its insides. If there is a chemical affinity between the elements of calcite (the oyster shell) and silica (the stone), there is also an improbable likeness on the level of avidity — a veritable craving for proliferation and difference, which is at the heart of things. We also perceive another image in this split stone: the moment of creation, the so-called big bang, the radiant folds of the universe stretching outward.

Conversely, too, the oyster becomes a stone by mimicking stone's textures. But ultimately the oyster only pretends. For its speeds of development and emergence are very different; an oyster breathes, moves, and grows much faster than a stone does (we are suspending the discussion here if stones can breathe). In addition, its secret, what lies inside of its calcite envelope, is different than the secret of the stone and much more vulnerable.

Being one of the most recognizable architectural facades, the baroque, either as a period or as a synecdochic adjective, is generally understood to refer to art, music, or architecture characterized by florid, convolute, ornamental decoration. But there is a more specific reason, besides the architecture of its façades, for evoking the baroque when looking at an oyster's interior. Etymologically, baroque comes from *barocco* (in French, Portuguese, Italian) meaning "irregular pearl." Hence the intermittent quality of its surface corresponds to the ornamental eccentricity of the baroque façade, while the intrinsic beauty manifested in its interior pearl corresponds to the spiritual value of the sacristy.

Furthermore, the image of the oyster becomes purely baroque if we consider the following definition:

> Baroque architecture can be defined by this severing of the façade from the inside, of the interior from the exterior, and the autonomy of the interior from the independence of the exterior, but in such conditions that each of the two terms thrusts the other forward.[2]

While there is a clear independence of the exterior shell (in terms of texture, hardness, chemistry, function, aesthetics) from the autonomy of the interior (as a filtering machine, food delicacy, supple amassing of the sea), they nevertheless thrust each other forward. The life of the oyster depends on the communication between the inside and the outside.

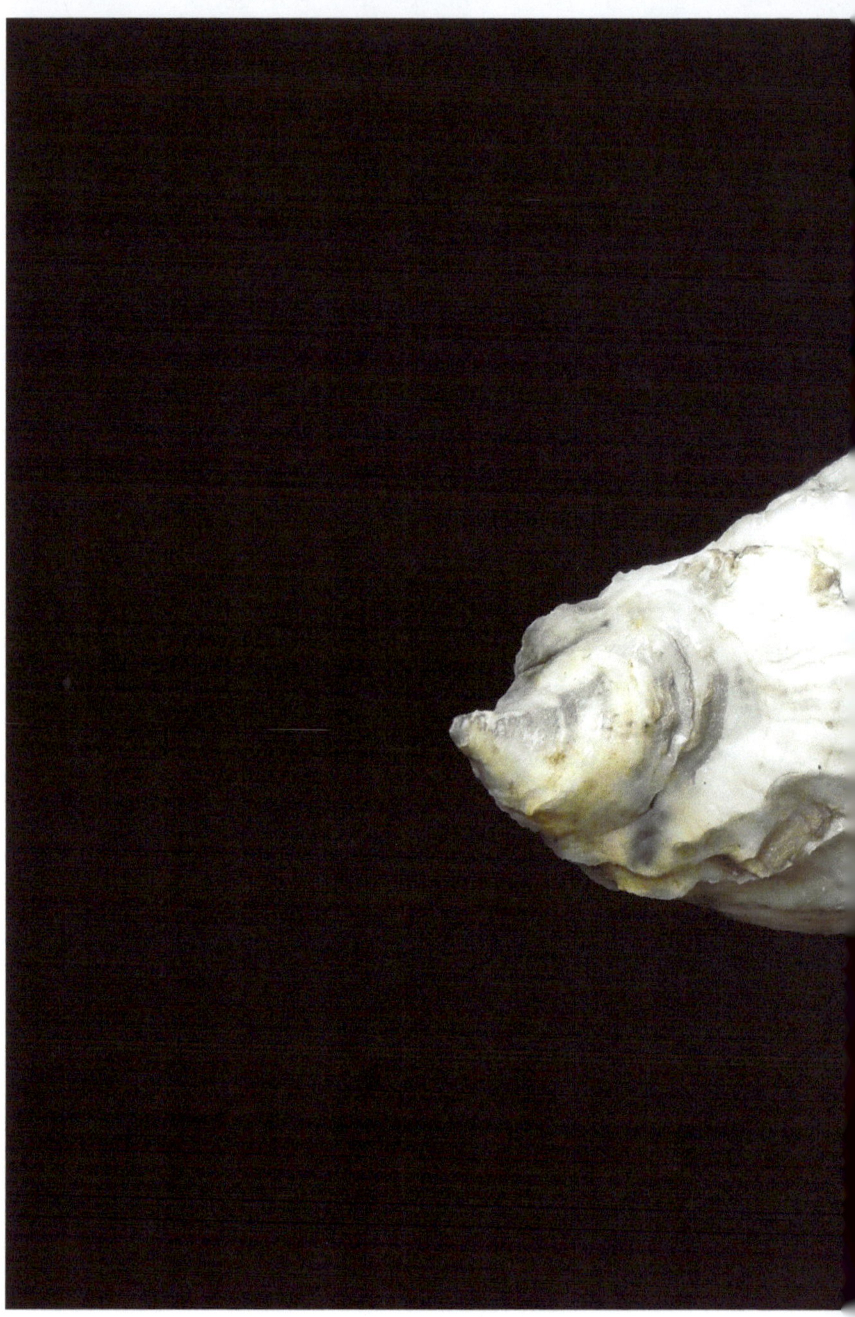

Fig. 2

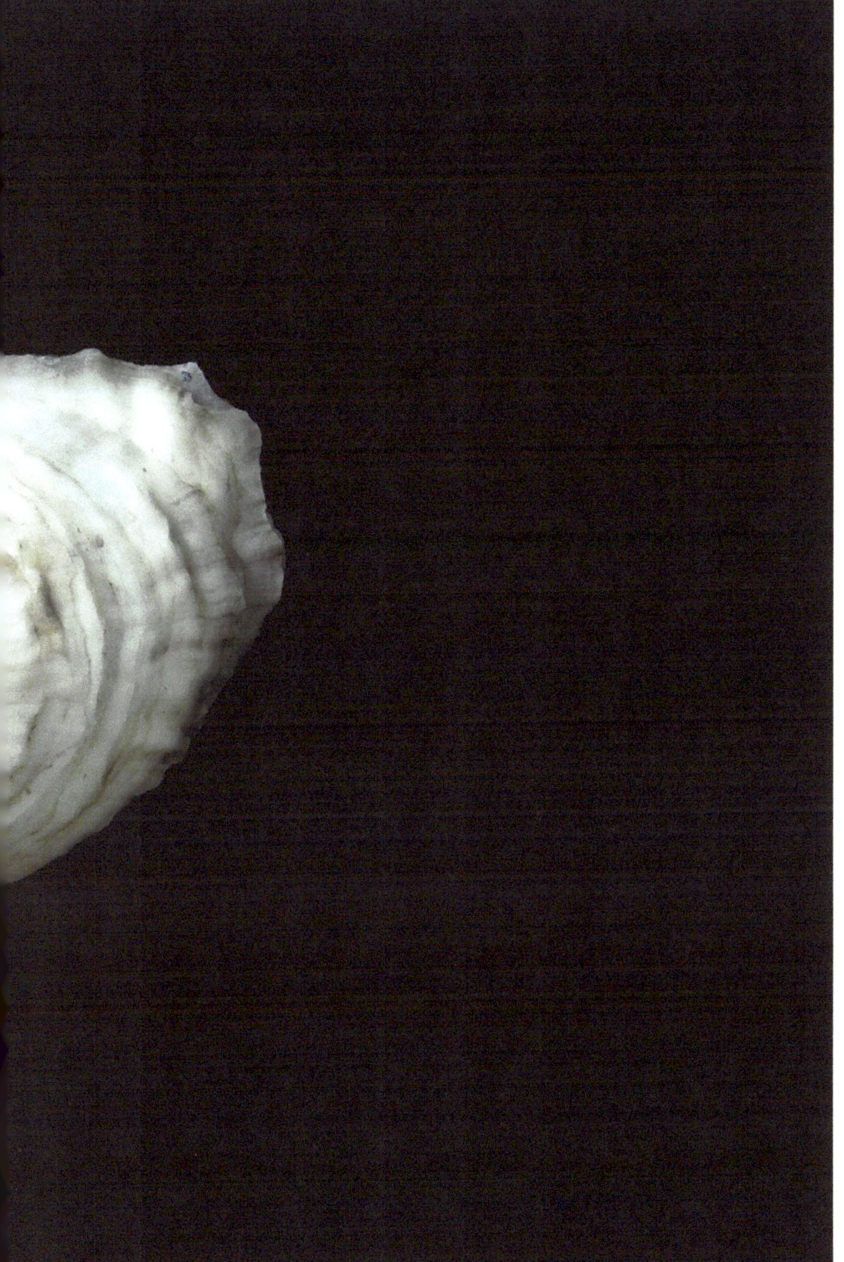

11. Baroque Fulcrum

Cosmic indeterminacy vs. contextual contingency. Which wins? →

The image of the shell thus reveals nothing short of an architectural edifice, a futuristic cathedral, or an ahistorical baroque expression, both familiar and unfamiliar in its curvature (Fig. 2). Except the former emerged in the seas of the Cambrian period 500 million years ago, while the latter occurred in 1600s Catholic Europe. The former relies on the geometry infused from deep cosmological time; the latter on the intricacies of Austrian or Italian mathematical calculations where a shift of one millimeter would make the whole interior falter.

We call it ahistorical because the problem of the inside and the outside far exceeds architectural or artistic time periods. On the other hand, one also has to see specifically baroque problems exceeding their own contexts and times, and seeping into other historical and geologic periods, albeit in a different configuration.

This exceeding of the inside/outside problem also occurs on the level of feelings:

> Perhaps that's what I feel, an outside and an inside and me in the middle, perhaps that's

> what I am, the thing that divides the world in two, on the one side the outside, on the other the inside, that can be as thin as foil, I'm neither one side nor the other, I'm in the middle, I'm the partition, I've two surfaces & no thickness, perhaps that's what I feel, myself vibrating, I'm the tympanum, on the one hand the mind, on the other the world, I don't belong to either.[3]

He is a bivalve that is a fulcrum, and then pure surfaces without thickness, then pure vibration, or an instrumental surface that reacts to vibrations, a tympanum that can be seen as a mantle ("as thin as foil"). But we should not really say "he," for there is no character left here, no human or animal, only the experience of an inside and an outside, and a non-belonging to either side. No character left here, only a filter.

The oyster does not resolve the problem of the inside and the outside either; rather, it exasperates it by making it visible. Through its baroqueness, the architecture of the façade protects the formless fulcrum. The outward coarseness is simply a series of complex folds generating the façade of the shell. But just as each façade is different, every oyster differs from the next. The notion of difference is essential here:

Obstacle becoming a prop; a hinge that makes action possible (e.g. a joint in a bird's wing, a lever for a boat's oar). →

> Difference is the alpha and omega of the universe; everything begins with difference, with the elements whose innate diversity (which various reasons make probable) can in my view be the only justification of their multiplicity; everything ends with difference, where, in the higher phenomena of thought and history, it finally breaks free of the narrow circles in which it had bound itself, namely the atomic vortex and the vital vortex, and transforming the very obstacle it faced into a fulcrum, surpasses and transfigures itself.[4]

This passage can itself be seen as a thought-oyster: an insular wall with enormous richness at its core. It tells us that difference is like an oyster: a filtering machine that turns the obstacle (the narrow circles of mechanized sameness) into a central support of its own transformation. Difference always overcomes itself. This is its law of spontaneity, both a rule & improvisation. Written 200 years earlier than the above passage, article 9 of Leibniz's "Monadology" proposes an identical idea:

17th century thought, forever co-present, eternal. →

> Each monad, indeed, must be different from every other. For there are never in nature two beings which are exactly alike, and in which it is not possible to find a difference either internal or based on an intrinsic principle.[5]

So it is with each oyster we take into our hand. It is born out of endless differentiation of itself.

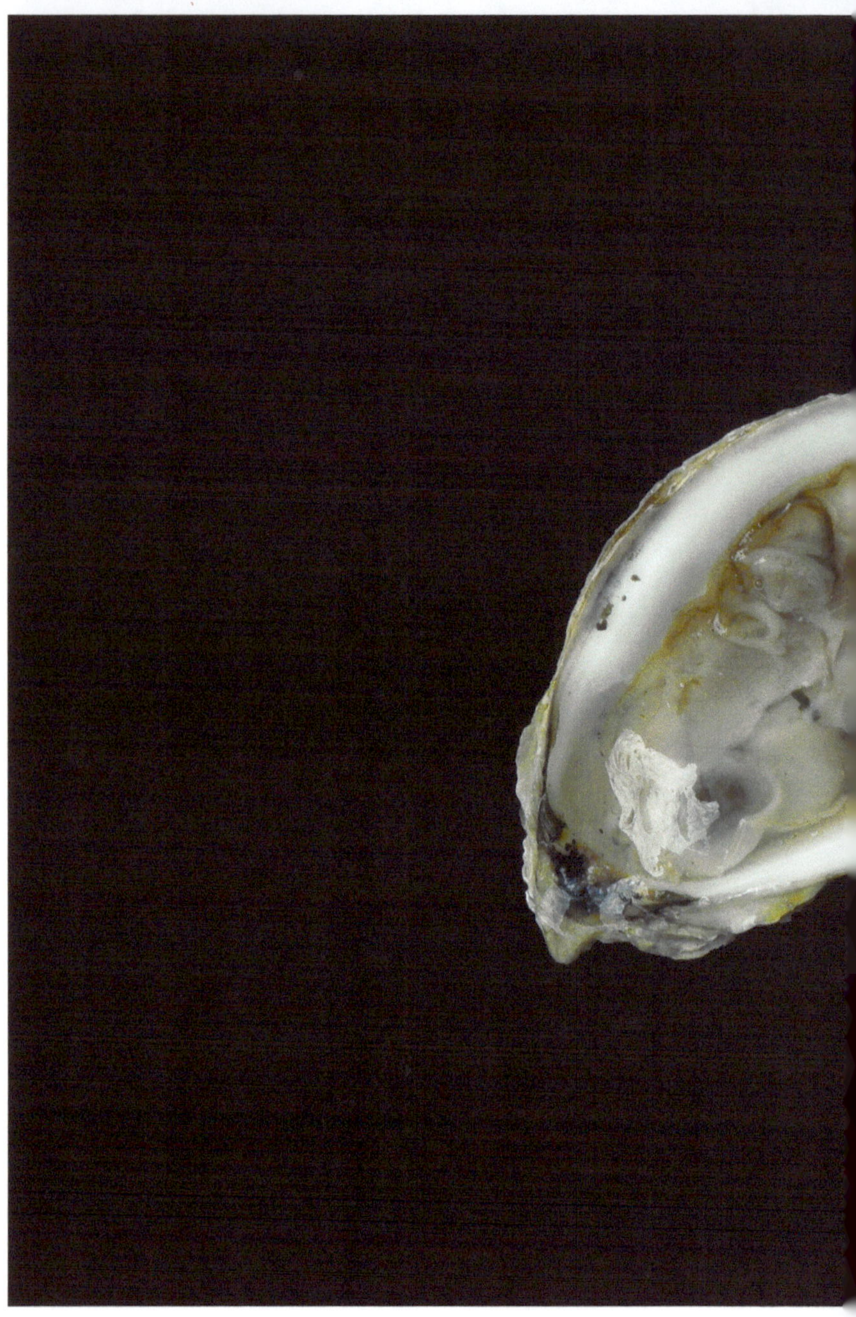

Fig. 3

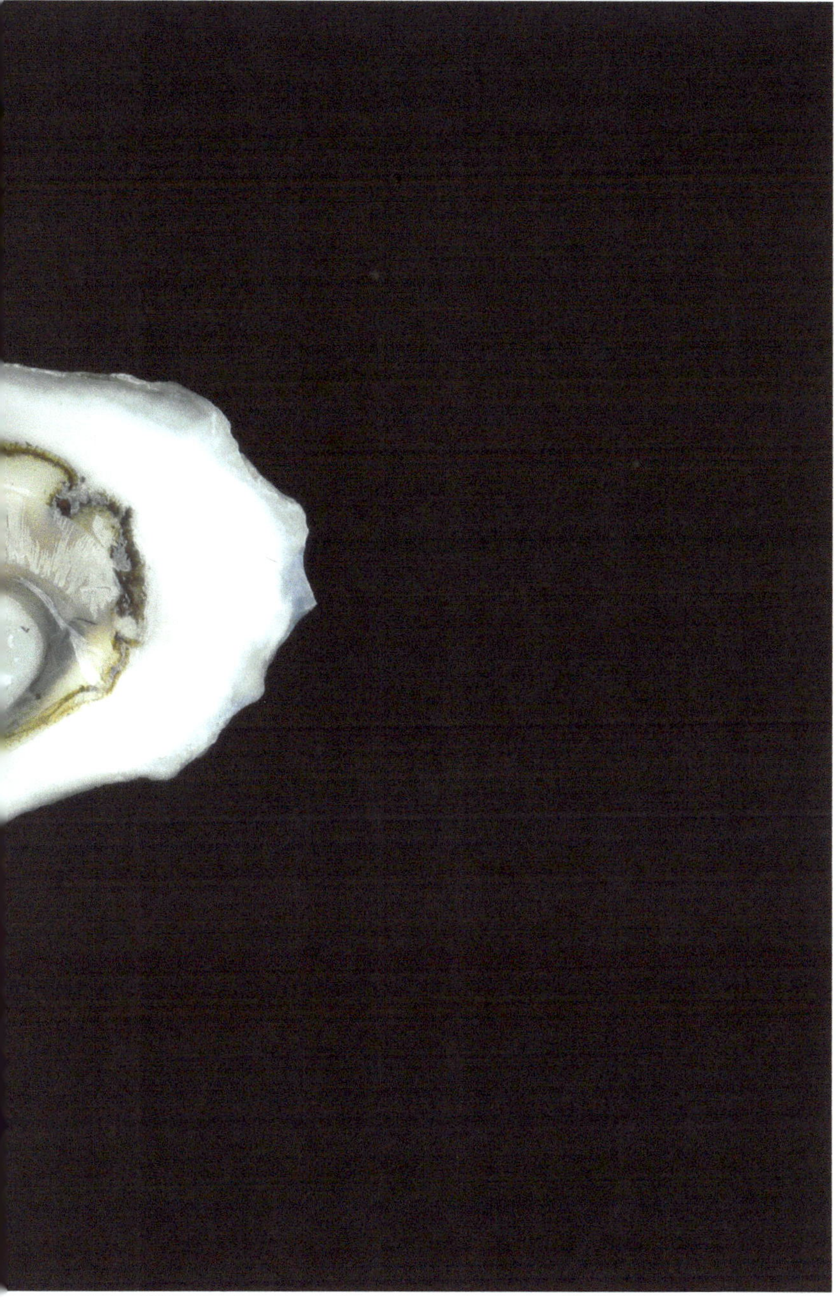

III. Primordial Soup

Forgotten, glorious, "diagonal science" of Roger Caillois, applied. →

Underneath the façade, or the shell, is an enveloping organ called the mantle: a dorsal body that is just another fold. The mantle is a skin that secretes calcium carbonate, which creates the shell, which in turn keeps the inside of the oyster secret (concealed). Moreover, the mantle is a cloak, a cape that in many mollusks extends into flaps. It is here that conchology (the science of shell formation) collides with architecture and fashion, as chemistry converges with aesthetics in the production of the cloak-fold.

The composition of the oyster can be said to contain an exo-dimension (the mantle and the shell) and an endo-dimension (the insides). More specifically, the inside of the oyster resembles a primordial soup; a little scoop of the essence of the sea (Fig. 3). Soup is a place where different ingredients lose their form.

A little game; order *&* chaos caressing each other. →

The truth is that chaos reigns inside the shell — an undifferentiated mass which is difficult to admire. It is the outside, the façade, that provides the sense of beauty because of its solidified permutations. A composition of folded thresholds, the oyster is a material manifestation of the process in which chaotic forces assume a harmonic state: from the chaotic soup continue the layers of the mantle and then of the shell.

Not only does the interior of the oyster, the secret universe of its inside, look like spit (easily evoked by its bubbles), but the oyster larvae are called "spat," a word that denotes pouring out & expanding (the spawn, the generation), & a past participle of the act of spitting. The spat condenses:

George Bataille's idea (image of cosmos as spit) amplified. ←

> Encountered a planet of baby oysters: spat. Half a million in a bundle the size of a human head. Each appears fully formed but about the size of a small pebble or a grain of rice. All of the oysters that I have seen here begin the same — it is only in relation to the waters where they grow that they become so multiplicitous. Divergent expansions. Where do they originate before this? Must be condensations in the mist. Fertile mixtures of warm and cool currents. Today is our largest day so far — over 40,000 oysters will come in and go out.[6]

Aligned with Leibniz & Tarde, unconsciously. ←

Images proliferate here: the spat is a planet, the planet is a human head, all originating out of a formlessness of mist. It is a clever way of assigning the generation of oysters to mist, molecular clusters of water. But if one looks very closely, each micro-droplet is already a tiny oyster, inside of which there is condensation, resulting in the primordial soup. Everything is within everything else, folded inside the fold.

Evolution and involution, expansion and contraction, are thus expressions of foaming rather than of bifurcation. The spit in this respect is a better image of evolution than a tree. Here harmony lies precisely in the undifferentiated matter that is always in the process of expanding (as an oyster never stops growing), bubbling (as the foam of the sea), swelling (as the diffusion of galaxies).

Harmony is not only the arrangement of forms; it is also the arrangement of correspondences between one thing and an other: "Consequently, every body responds to all that happens in the universe, so that he who saw all could read in each one what is happening everywhere, and even what has happened & what will happen."[7] All things conspire: the oyster and the plankton and the sea.

An Oyster, cracked open

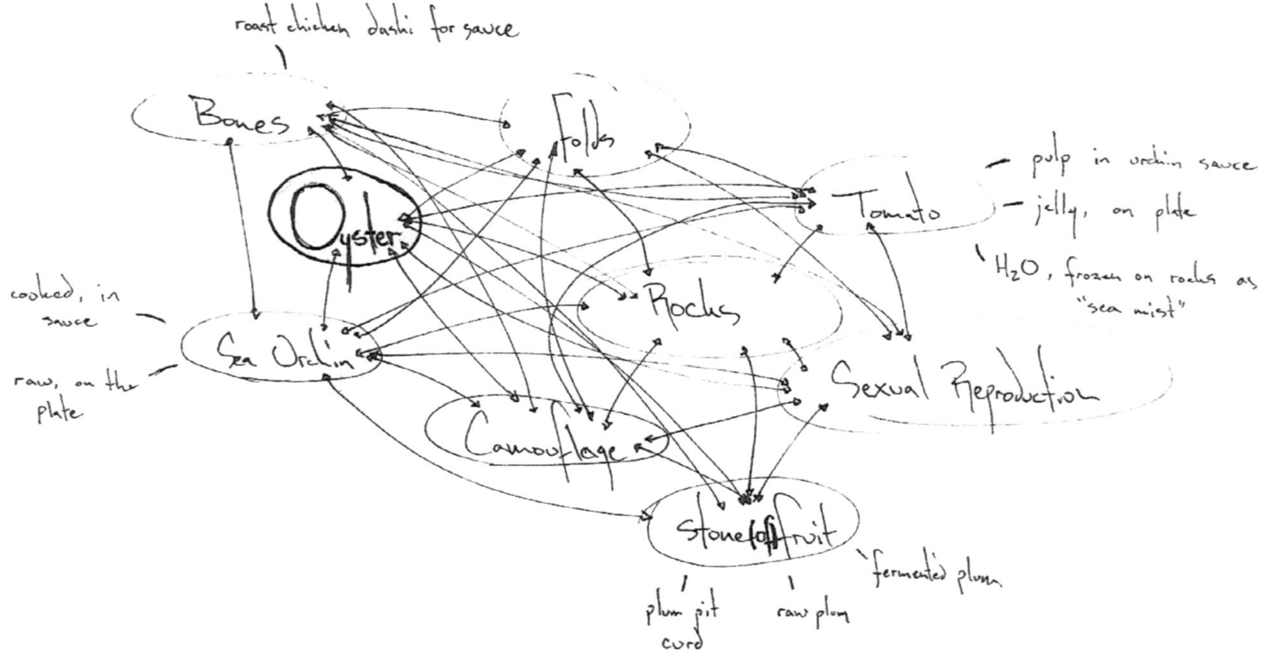

roast chicken dashi for sauce

Bones

Folds

Oyster

Tomato
- pulp in urchin sauce
- jelly, on plate
- H_2O, frozen on rocks as "sea mist"

Sea Urchin
- cooked, in sauce
- raw, on the plate

Rocks

Sexual Reproduction

Camouflage

Stone (pit) fruit
- plum pit curd
- raw plum
- fermented plum

fermented plum
plum pit curd
raw plum
sea urchin
tomato jelly
oysters, kissed by the grill
urchin sauce
rocks with "sea mist"
filtration
earth
ocean
sex
absorbtion
territory
excretion

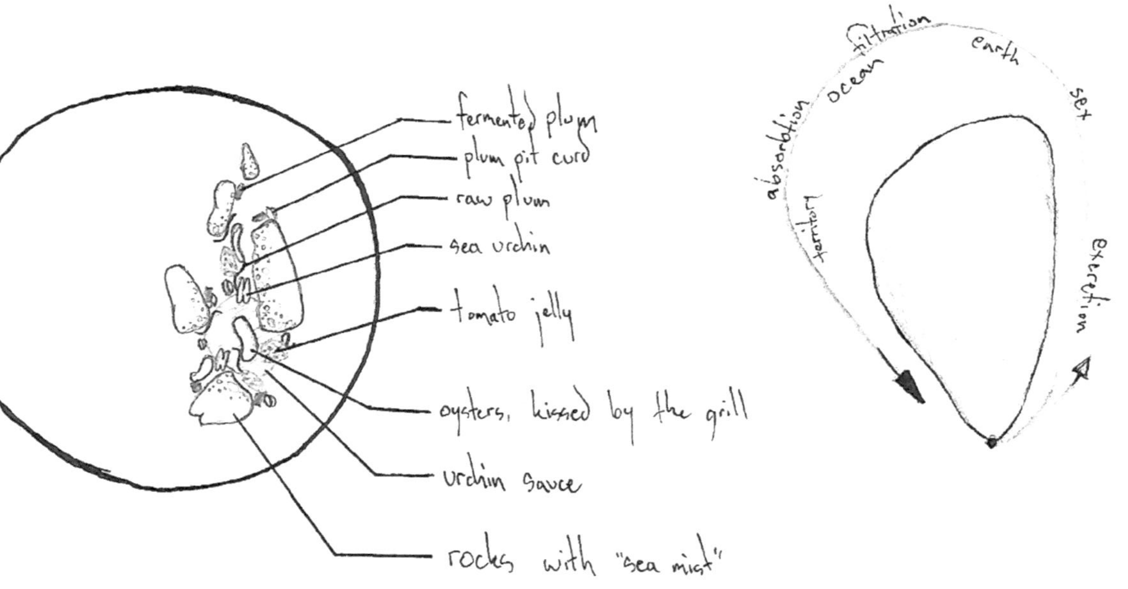

Fig. 4

IV. Cracked Open

In addition to its architectural æsthetics of the outside shell, the skills to become a stone, and the ability to produce a pearl, the oyster is also food. Upon forcibly opening it, all that is contained within it is a small ugly lump, as unrecognizable as any living thing. It has no mouth, no eyes, no definable organs — it is just a folded chunk of organic matter.

Thinking through taste buds; tongue-thought. →

We eat the formlessness, ambiguously looking at it with fascination and disgust before we swallow it. There is no cutting; we slurp it whole. A microcosm of the world absorbed just as the thing itself absorbed its entire diet. The flavor and texture is indivisible; it cannot be described any better than as approximately the taste of an ocean and all of its scents, flavors, and denizens condensed into a slimy mass.

Waking the dormant element, superior culinary tactic. →

A young chef knows how to intensify a hidden element that is waiting to be released, that is permanently brewing in its substance (a dark element that initiates its growth). Through cruelty (dismembering oyster parts) he extends the latent potentials of the oyster-substance. In this inventive dish diagram (Fig. 4), all the modalities of the oyster, organic and inorganic, are taken into account. The rock is tied to the plum's pit (they share the same hardness), camouflage to the sea urchin's rawness, sea mist to sexual reproduction. As all ingredients are positioned on a circular plate, like an archipelago of life itself, they become a meal. In the process, the chef also sees the border of the oyster's shell enlarged; he creates another encircling line where acts of filtration, absorption, and exertion swell the actual lines of the shell.

Kissed by the burning grill, the oyster exceeds itself. On the plate, the affinities between the ingredients (substances) form an archipelagic geometry.

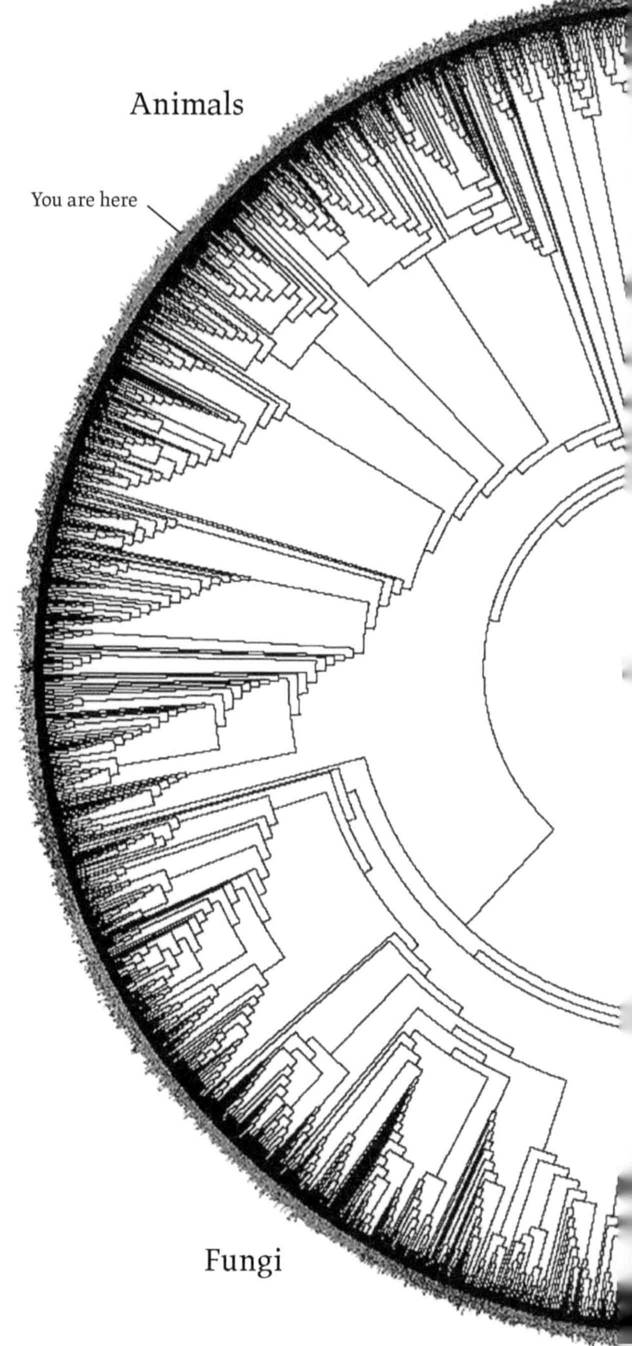

Fig. 5

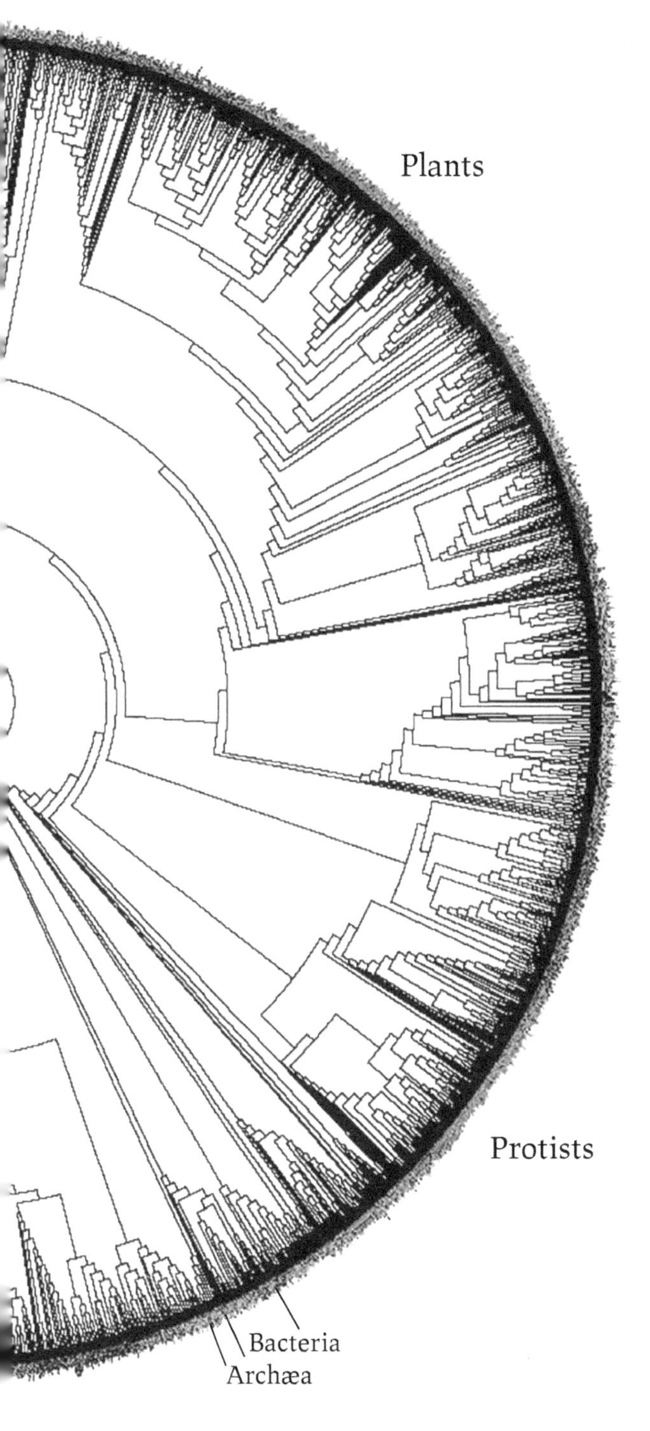
Plants
Protists
Bacteria
Archæa

v. Clandestine Note

If it is not an animal, what is it? There seems to be an error in the taxonomic system for there is nowhere else to put the oyster.

When it comes to contemporary classification systems, biologists are moving into a rhizomatic model called phylogenetics, which are represented by cladograms (Fig. 5). The cladogram is an alternative form of diagraming a traditional taxonomic tree which makes the nonhierarchical unfolding of principles of life more visible. For example, the image that represents the family tree of all species can be described as:

> a circular representation of the tree of life, in which time starts at the center, roughly 3.5 billion years ago, and proceeds outward to the present. The full resolution image contains 3000 species, out of the approximately 9 million species thought to exist.[8]

Another dish diagram? An entire evolutionary creation as a meal, of one another... →

In line with this, there is a strange, clandestine note written by the most famous representative of the tree of life model, which undermines its own authority. Darwin: "The tree of life should perhaps be called the coral of life, base of branches dead; so that passages cannot be seen."[9] And in another place: "Tree not good simile. [Instead it would be better to say] piece of seaweed endlessly branching."[10] Evolution of life should then be depicted as a jumble of curvatures, a chaotic harmony of relations, falling arcs of spontaneous initiatives. Not really species, then, but rather aggregates of sieving.

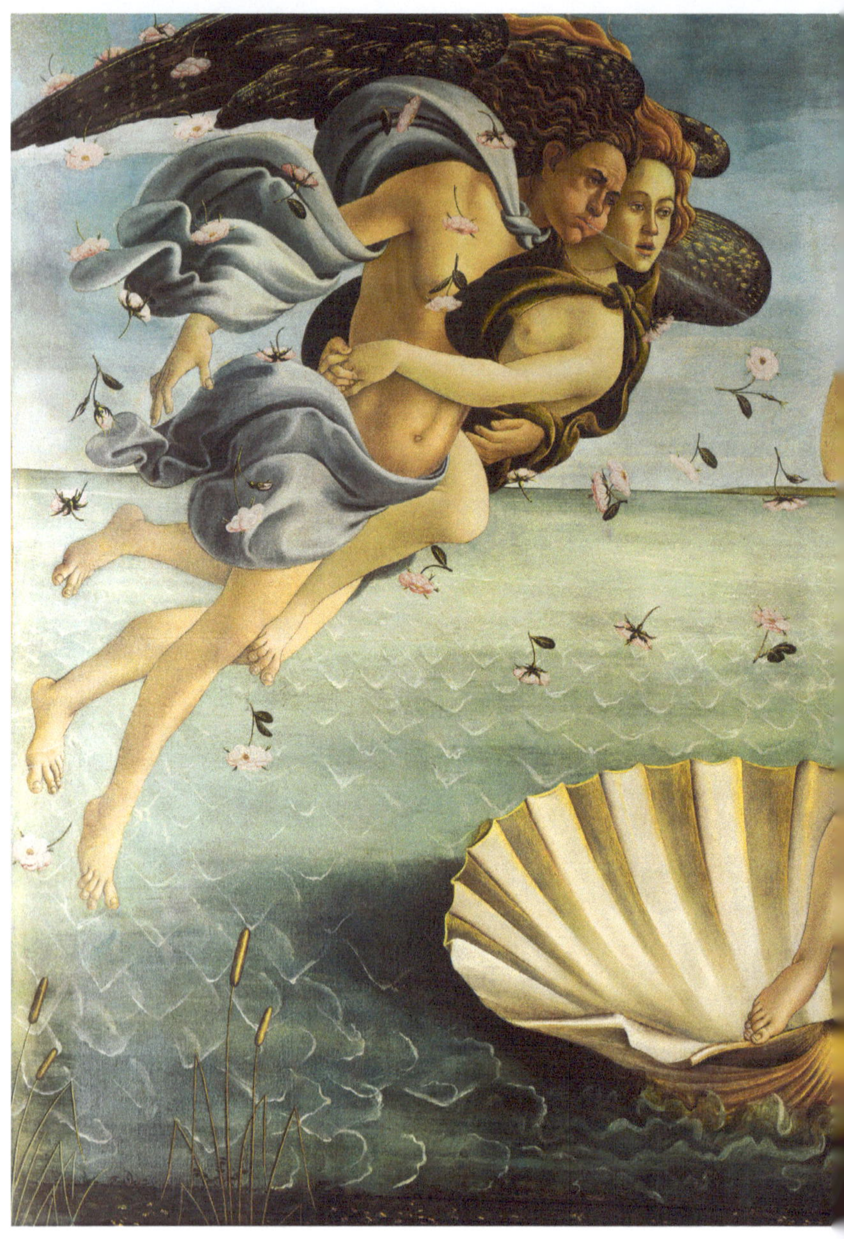

Fig. 6

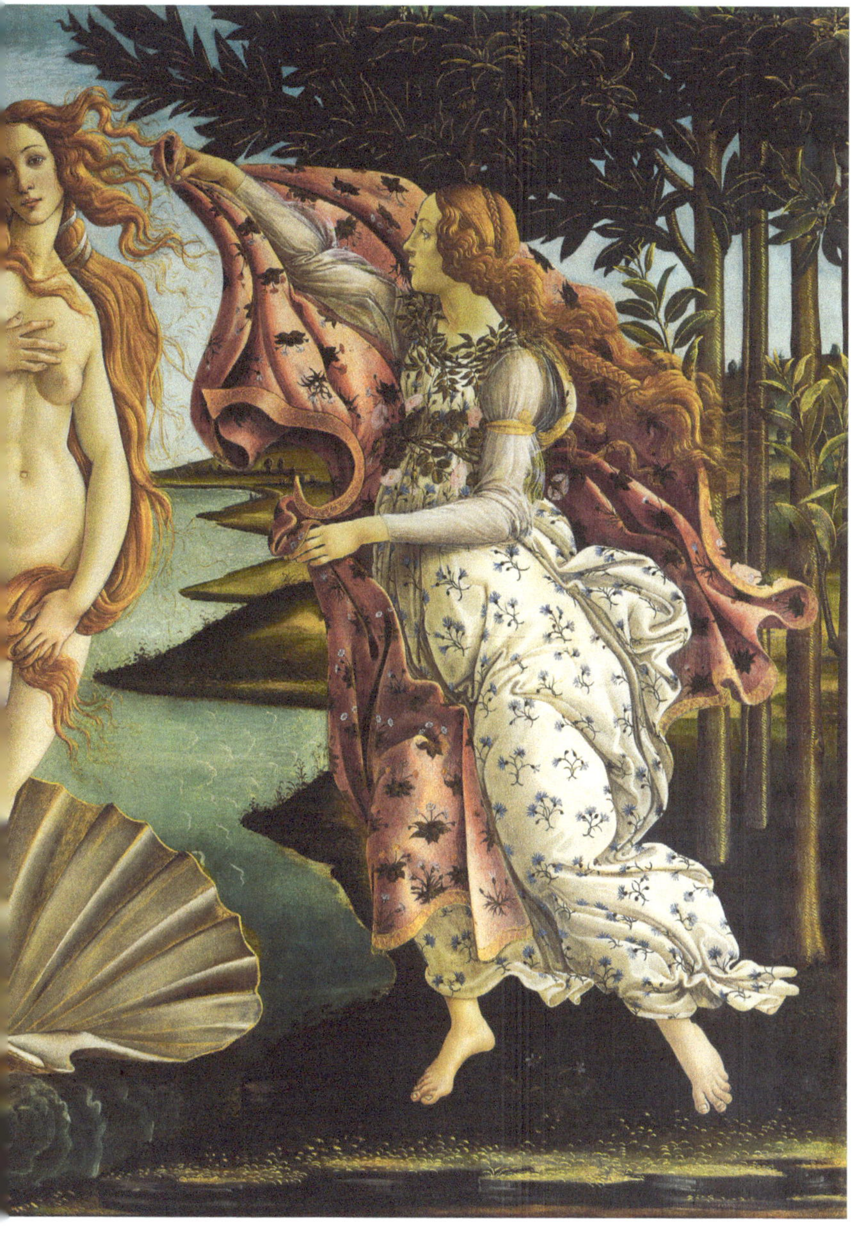

VI. Venereal Seduction

It is no coincidence that bivalves are a powerful symbol of fertility in art. It is often posited that this comes from the bivalve as a vaginal image; however, there is a more compelling understanding of this figure. We see it in the famous Botticelli painting *The Birth of Venus.*

Here, Venus, the goddess of fertility, presents herself as standing, floating, seductively, on the inside of a shell (Fig. 6). Except for the rigid trees, it is folds that surround her appearance: of her hair, of the clothes of the figures that approach her, of rippling waves, of angels' wings, of angels' breath, of the mantle that is being thrown at her, of the curving outlines of the coast. Is she the pearl?

When it comes to the birth of seduction the lesson reads: out of the slime and spit comes the pearl. At the core of the oyster, the ugly lump, lies the epitome of beauty. The power of the pearl is properly venereal, for it is an object of seduction, and the object of perennial attraction. This little adornment, as if it came from the mouth of Venus herself (the crystallized saliva of the goddess), is in radical economic and aesthetic disproportion with the oyster — the machine that gave it life. For despite all its exceptionality, its philosophical and alimentary richness, the oyster is at the same time devastatingly cheap: one dollar for one oyster.

VII. Energetic Verve

Returning to the primordial dimension of the oyster, we see it as the intersection of mineral, amœba, plant, and animal. All microscopic ocean stuff, such as algæ, plankton, and other organic particles that are believed to be the earliest life forms, are present in the oyster, both conceptually and literally. But they are also the oyster's nutrition, and they can move, but it cannot. Since it filter-feeds, the oyster survives off these basic forms of oceanic life. Since these particles form the basis of terrestrial life, in a very real sense, all of life is contained in the sachet that is revealed. It sits and waits and allows them to pass into its shell where it uses them for nutrition. The oyster absorbs the world.

All classifications are childish; scales interpenetrate, vast and minuscule. →

The shell and the static nature of the oyster links it equally to the mineral. Being at this bizarre intersection of nature gives the oyster its ardor, one that ruptures not only the traditional classification of organisms, but also the clear-cut differentiations between the mineral and organic spheres. Francis Ponge understood this low energetic verve well when he wrote the prose poem titled "The Oyster." He first points to the mineral aspect of the oyster, then extends it to the mystery of the interior and the pleasure of opening it. He even describes the oyster as its own world, stubbornly "closed in upon itself" (a monad):

> Roughly the size of a rather large pebble, the oyster is more gnarled in appearance, less uniform in color, and brilliantly whitish. It is a world categorically closed in upon itself. And yet it can be opened: that takes gripping it in a folded rag, plying a nicked and dull-edged knife, chipping away at it over and over. Probing fingers get cut on it, nails get broken. It's a rough job. The pounding you give it scars the envelope with white rings, a sort of halo.
>
> Within, one finds a world of possibilities for food and drink: beneath a mother-of-pearl firmament (strictly speaking), the skies above settle in on the skies below, leaving only a rock-pool, a viscous greenish sack that ebbs and flows before the eyes and nose, fringed with a border of darkish lace.
>
> On rare occasion the perfect formula pearls up in its nacreous throat, and we take it at once for our adornment.[11]

Fig. 7

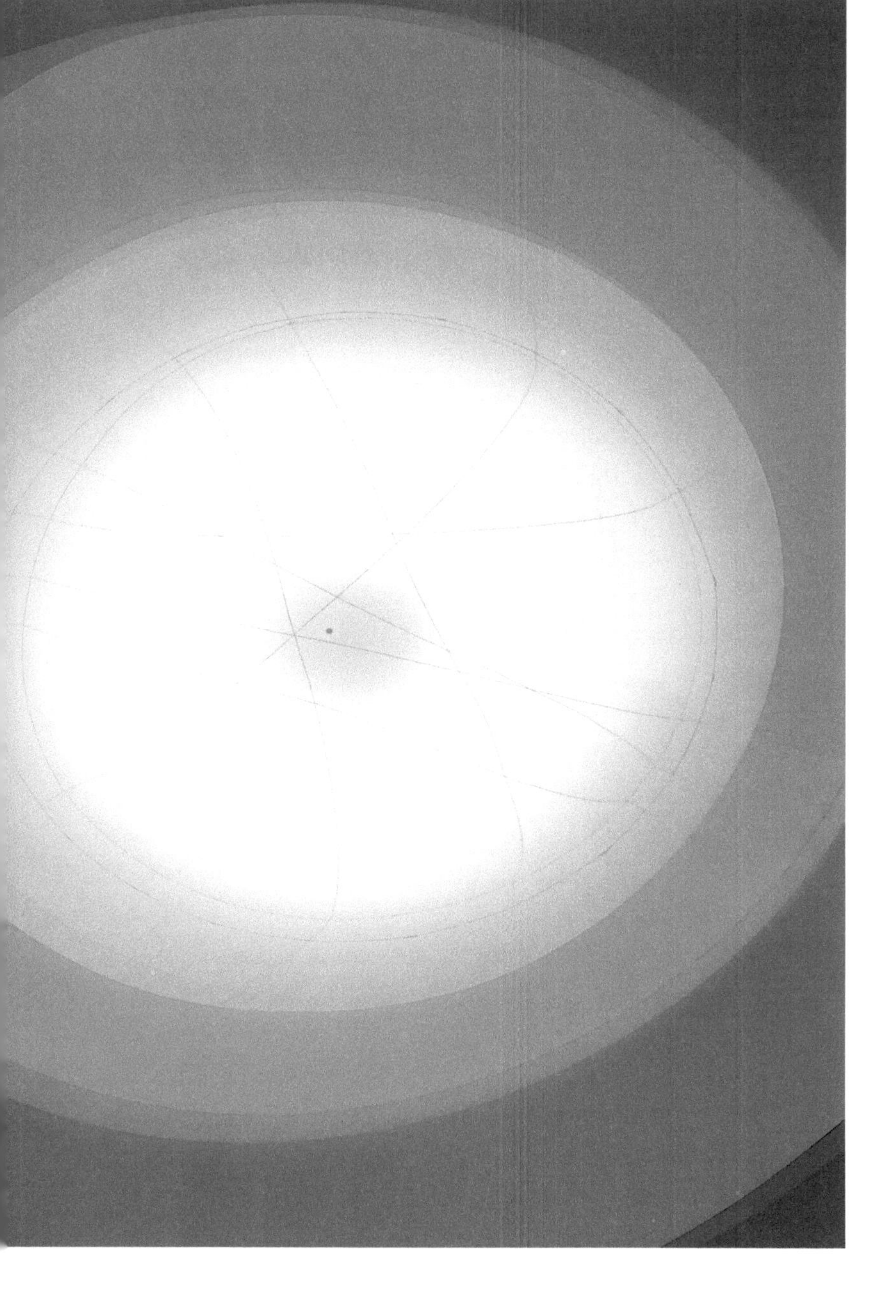

VIII. Secretion

In this absorption of the world, something else occurs, something both ordinary and extraordinary. When a parasite, a foreign body, finds its way to the mantle of the oyster, "the mantle secretes nacre, or mother-of-pearl, by synthesizing calcium carbonate from materials in the water."[12] The pearl sac enfolds the parasite, entombs it. A pearl is nothing but a discrete and indivisible unit formed by time and pressure within an oyster. The perfect sphere is thus secreted into the world, as a revolt inside the oyster. In this sense, life is revolutionary as long as it produces variations based on an internal conflict.

Beauty, adornment, is chemical. →

And political. →

We devour the oyster and exhibit the pearl, eradicating the formlessness and memorializing the spherical perfection. One should recognize this need for forms as both a legislative aspect (the impulse for rules and regulations) and a theological one (the impulses for preservation and admiration). Hence Peter Sloterdijk's statement: "Worldly spherology is the attempt to free the pearl from the theological oyster."[13] However, this insight has to be reversed, for it is the pearl that is the theological (Venus herself), not the oyster. Speaking crudely, the pearl is the secretion of a mollusk, a crystallization of its folds and bubbles, just as any institution is the crystallization of an idea (e.g. the democratic state is the pearl of the idea of democracy).

The pearl is the crystallization of power, the authoritative form out of which values of prestige emerge. And yet what is in the pearl that makes it into a valued harmonious form? What struggles independently from the formlessness in which it is born? What procedures of chaos does it summon into its perfection? The pearl is not an end point; something else wills its existence from the inside. For already in the pearl there is something revolting against it (there is always the chaotic fulcrum), a rebellion at the core of the crystalized rule.

Free-will. ←

The oyster emerges under the waves, the constant rhythmic folding & unfolding of the water. It is the hardening compound of the water. Chemistry points toward vitality, and the vital series goes further than chemical reductions; it ties the oyster to the wave, the wave to the sea, sea to spit, spit to universe, and universe to pearl.

Philosopher Otto Bollnow said that the baroque is paradoxically determined by an architecture of unlimited interior, which always strives toward infinity, by breaking space with optical illusions and running through the ceiling.

This is precisely how we can define the interior space of Christo's *Big Air Package* installation (Fig. 7), which is the inside of a gasometer, a former gas container where natural gas lived under atmospheric pressures in ambient temperatures. If we stare into the image (a snapshot of its ceiling), we see the pearl in the middle of it: a materialized, concretized light out of which radiate layers as protective mantles that muffle its brightness. This pearl-sphere is actually an opening through which light from the outside enters, but the optical illusion of space is complete. Look closer: a secretion of light geometricizes the infinite which contains some throbbing dark element within its core. It is a thing that filters, and which we will swallow. It will either nourish or poison.

Endnotes

1 See Gilles Deleuze, *The Fold: Leibniz and The Baroque*, tr. Tom Conley (1993) 28, where he defines a philosophical monad as a sacristy.
2 Ibid., 28.
3 Samuel Beckett, *The Unnamable* (1958) 134.
4 Gabriel Tarde, *Monadology and Sociology*, ed. & tr. Theo Lorenc (2012) 40–41.
5 Leibniz, *Monadology and Other Philosophical Essays* (1965) 252.
6 Will Scarlett, personal email.
7 Leibniz, *op. cit.*, 264–265.
8 `http://fractalfoundation.org/OFC/OFC-8-4.html`
9 Charles Darwin, *Notebooks, 1836–1844: Geology, Transmutation of Species, Metaphysical Enquiries* (1987). We are grateful to Sam Yehros for sharing this incredible discovery.
10 Quoted in H.E. Gruber & Katja Bödeker (eds), *Creativity, Psychology, and the History of Science* (2005) 263.
11 Francis Ponge, *The Nature of Things*, tr. Lee Fahnestock (2011) 22.
12 "Oysters: Gem of the Ocean" (`http://www.economist.com/node/12795573`)
13 Peter Sloterdijk, *Bubbles: Spheres I*, tr. Wieland Hoban (2012) 54.

Image Credits

Fig. 1 Roger Caillois, *The Writing of Stones*, chalcedony, Rio Grande do Sul, Brazil; 94 × 92 mm.

Fig. 2 Will Scarlett, *Singularities* (*Oyster Series*) (2013).

Fig. 3 Will Scarlett, *Singularities II* (*Oyster Series*) (2013).

Fig. 4 Joshua McCarty, *An Oyster, Cracked Open* (2014). Diagram for a meal.

Fig. 5 A cladogram showing divisions and convergences of life.

Fig. 6 Sandro Botticelli, *The Birth of Venus* (c. 1486).

Fig. 7 Christo, *Big Air Package, Gasometer Oberhausen*, Germany (2010–13). Photo: Wolfgang Volz / laif / Redux.

Acknowledgements

It took considerable time and energy and affirmation for this little molluskular text to be born. First, in 2013, Nik Kosieradzki provided the kernel that connected the marine creature to the baroque. This was for the "Ethnographic Image" class at Reed College. I augmented it by providing layers, one upon another, from the most limber to the most firm. Craig Epplin helped with initial editorial formatting. Later, I pried it open again and discarded much debris. Then I let it mature and grow in darkness. Finally, Emile Plateau, the curator-chef of Contra Mundum Press, provided a perfect environment, with expansive and smooth sheets, nourished by typographer Alessando Segalini and visual artist Federico Gori, like the three layers of the Sea (of the surface, of depth, and of the seafloor sediments). Each of these is present, with its own temperature, in the text. This work of calcareous thought and supple geometry could not come to life, in this form, anywhere else. Now it lives only for the feast, for the self-expenditure, which is yet to come.

—DL

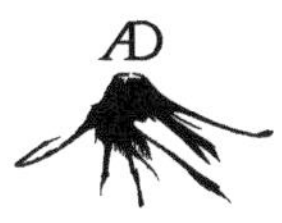

The Contra Mundum Press Agrodolce Series

Agrodolce, the sweet & sour, two opposite elements that create a tangy contrast, evoke the agon, *Innigkeit*, the *coincidentia oppositorum*, unsuspected & surprising amalgamations. Spurred by this, one thinks of the two primary opposing elements of the universe, as conceived by Empedocles: love and strife. The philosopher that hails out of Akragas (Agrigento), Sicilia, the surrogate birthplace of philosophy, the philosopher's true home. Separate from any mainland, separate from nationalism. A culture born of centuries & centuries of amalgamations. More exigently, its mythic site is Mt. Etna, the volcano that, as legend has it, Empedocles leapt into. Etna, the agricultural home of apiaries, vineyards, olive and citrus groves, pistachio and hazel trees, wild herbs and far more. In erupting over 12 times a year, the landscape of Etna is one that undergoes continual metamorphosis, with eternally-evolving soil comprised of basalt, pumice, and ash. The Agrodolce Series, a source of unorthodox books on food.

Curated by Emile Plateau

COLOPHON

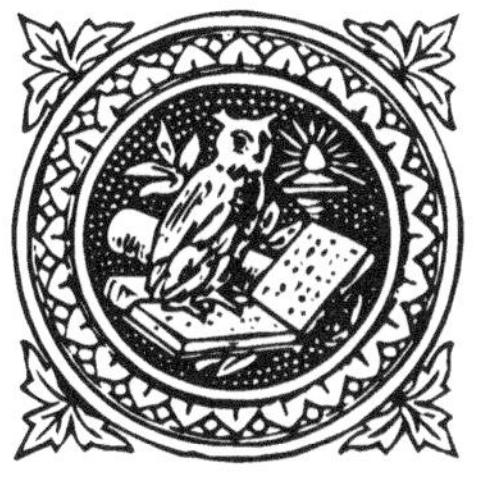

THE OYSTER
was handset in InDesign CC.

The text typeface is *JAF Lapture*
The display typeface is *Adobe Warnock*

Book design & typesetting: Alessandro Segalini
Cover image: Federico Gori
Cover design: CMP

THE OYSTER
is published by Contra Mundum Press.

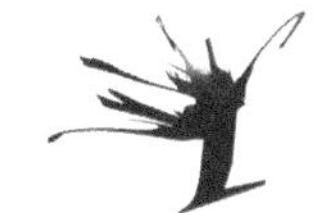

Contra Mundum Press New York · London · Melbourne

CONTRA MUNDUM PRESS

Dedicated to the value & the indispensable importance of the individual voice, to works that test the boundaries of thought & experience.

The primary aim of Contra Mundum is to publish translations of writers who in their use of form and style are *à rebours*, or who deviate significantly from more programmatic & spurious forms of experimentation. Such writing attests to the volatile nature of modernism. Our preference is for works that have not yet been translated into English, are out of print, or are poorly translated, for writers whose thinking & æsthetics are in opposition to timely or mainstream currents of thought, value systems, or moralities. We also reprint obscure and out-of-print works we consider significant but which have been forgotten, neglected, or overshadowed.

There are many works of fundamental significance to *Weltliteratur* (& *Weltkultur*) that still remain in relative oblivion, works that alter and disrupt standard circuits of thought — these warrant being encountered by the world at large. It is our aim to render them more visible.

For the complete list of forthcoming publications, please visit our website. To be added to our mailing list, send your name and email address to: info@contramundum.net

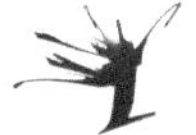

Contra Mundum Press
P.O. Box 1326
New York, NY 10276
USA

OTHER CONTRA MUNDUM PRESS TITLES

2012 *Gilgamesh*
Ghérasim Luca, *Self-Shadowing Prey*
Rainer J. Hanshe, *The Abdication*
Walter Jackson Bate, *Negative Capability*
Miklós Szentkuthy, *Marginalia on Casanova*
Fernando Pessoa, *Philosophical Essays*
2013 Elio Petri, *Writings on Cinema & Life*
Friedrich Nietzsche, *The Greek Music Drama*
Richard Foreman, *Plays with Films*
Louis-Auguste Blanqui, *Eternity by the Stars*
Miklós Szentkuthy, *Towards the One & Only Metaphor*
Josef Winkler, *When the Time Comes*
2014 William Wordsworth, *Fragments*
Josef Winkler, *Natura Morta*
Fernando Pessoa, *The Transformation Book*
Emilio Villa, *The Selected Poetry of Emilio Villa*
Robert Kelly, *A Voice Full of Cities*
Pier Paolo Pasolini, *The Divine Mimesis*
Miklós Szentkuthy, *Prae, Vol. 1*
2015 Federico Fellini, *Making a Film*
Robert Musil, *Thought Flights*
Sándor Tar, *Our Street*
Lorand Gaspar, *Earth Absolute*
Josef Winkler, *The Graveyard of Bitter Oranges*
Ferit Edgü, *Noone*
Jean-Jacques Rousseau, *Narcissus*
Ahmad Shamlu, *Born Upon the Dark Spear*

2016 Jean-Luc Godard, *Phrases*
Otto Dix, *Letters, Vol. 1*
Maura Del Serra, *Ladder of Oaths*
Pierre Senges, *The Major Refutation*
Charles Baudelaire, *My Heart Laid Bare & Other Texts*
2017 Joseph Kessel, *Army of Shadows*
Rainer J. Hanshe & Federico Gori, *Shattering the Muses*
Gérard Depardieu, *Innocent*
Claude Mouchard, *Entangled — Papers! — Notes*
2018 Miklós Szentkuthy, *Black Renaissance*
Adonis & Pierre Joris, *Conversations in the Pyrenees*
2019 Charles Baudelaire, *Belgium Stripped Bare*
Robert Musil, *Unions*
Iceberg Slim, *Night Train to Sugar Hill*
Marquis de Sade, *Aline & Valcour*
2020 *A City Full of Voices: Essays on the Work of Robert Kelly*
Rédoine Faïd, *Outlaw*
Carmelo Bene, *I Appeared to the Madonna*
Paul Celan, *Microliths They Are, Little Stones*
Zsuzsa Selyem, *It's Raining in Moscow*
Bérengère Viennot, *Trumpspeak*
Robert Musil, *Theater Symptoms*
Miklós Szentkuthy, *Chapter On Love*

SOME FORTHCOMING TITLES

Marguerite Duras, *The Darkroom*
Hans Henny Jahn, *Perrudja*

THE FUTURE OF KULCHUR
A PATRONAGE PROJECT

LEND CONTRA MUNDUM PRESS (CMP) YOUR SUPPORT

With bookstores and presses around the world struggling to survive, and many actually closing, we are forming this patronage project as a means for establishing a continuous & stable foundation to safeguard our longevity. Through this patronage project we would be able to remain free of having to rely upon government support &/or other official funding bodies, not to speak of their timelines & impositions. It would also free CMP from suffering the vagaries of the publishing industry, as well as the risk of submitting to commercial pressures in order to persist, thereby potentially compromising the integrity of our catalog.

CAN YOU SACRIFICE $10 A WEEK FOR KULCHUR?

For the equivalent of merely 2–3 coffees a week, you can help sustain CMP and contribute to the future of kulchur. To participate in our patronage program we are asking individuals to donate $500 per year, which amounts to $42/month, or $10/week. Larger donations are of course welcome and beneficial. All donations are tax-deductible through our fiscal sponsor Fractured Atlas. If preferred, donations can be made in two installments. We are seeking a minimum of 300 patrons per year and would like for them to commit to giving the above amount for a period of three years.

WHAT WE OFFER

Part tax-deductible donation, part exchange, for your contribution you will receive every CMP book published during the patronage period as well as 20 books from our back catalog. When possible, signed or limited editions of books will be offered as well.

WHAT WILL CMP DO WITH YOUR CONTRIBUTIONS?

Your contribution will help with basic general operating expenses, yearly production expenses (book printing, warehouse & catalog fees, etc.), advertising & outreach, and editorial, proofreading, translation, typography, design and copyright fees. Funds may also be used for participating in book fairs and staging events. Additionally, we hope to rebuild the *Hyperion* section of the website in order to modernize it.

From Pericles to Mæcenas & the Renaissance patrons, it is the magnanimity of such individuals that have helped the arts to flourish. Be a part of helping your kulchur flourish; be a part of history.

HOW

To lend your support & become a patron, please visit the subscription page of our website: contramundum.net/subscription

For any questions, write us at: info@contramundum.net

www.ingramcontent.com/pod-product-compliance
Lightning Source LLC
LaVergne TN
LVHW052309100826
845147LV00006B/709

9781940625423